DESERT

by Sheila Rivera

first step nonfiction

Lerner Publications Company · Minneapolis

What is a **desert?**

It is a place that gets very little rain.

Some deserts are cold.

Most deserts are hot.

A desert is a kind of
habitat.

A habitat is where plants
and animals live.

Trees grow in the desert.

Cactuses grow in the desert too.

Snakes live in the desert.

Owls live in the desert.

Foxes live in the desert.

This spider spins its **web** in the desert.

This mountain lion hunts for food.

A lizard warms itself on
a rock.

Camels walk across
the **sand.**

Many plants and animals
live in the desert.

Earth's Deserts

Desert

Desert Facts

 Most deserts are covered in sand.

 Some deserts are hot during the day and cold at night.

 Many deserts get less than ten inches of rain per year.

 The Atacama Desert in Chile receives less than ½ inch of rain a year!

Camels are the largest desert animals.

 Camels can go for long periods of time without water.

 Jackrabbits lose body heat through their huge ears. This helps keep them cool.

 Some desert animals hide in holes when the temperature is too hot.

Glossary

 cactuses – plants with thick stems and no leaves that live in hot, dry places

 desert – a dry place where plants and animals live

 habitat – where plants and animals live

 sand – small bits of rock

 web – a net of fine, silky threads spun by a spider

Index

The photographs in this book are reproduced through the courtesy of: © James P. Rowan, front cover, pp. 2, 5, 6, 8, 15, 22 (second from top, middle); © PhotoDisc Royalty Free by Getty Images, pp. 3, 12; © Karlene Schwartz, pp. 4, 9, 22 (top); © Tom and Pat Leeson, pp. 7, 17; © Michael & Patricia Fogden/CORBIS, p. 10; © Gary Schultz, p. 11; © Karen Tweedy-Holmes/CORBIS, pp. 13, 22 (bottom); © George H. H. Huey/CORBIS, p. 14; © Richard T. Nowitz/CORBIS, pp.16, 22 (second from bottom).

Map on pages 18–19 by Laura Westlund.

Lerner Publications Company
A division of Lerner Publishing Group
241 First Avenue North
Minneapolis, MN 55401 U.S.A.

Website address: www.lernerbooks.com

Library of Congress Cataloging-in-Publication Data

Rivera, Sheila, 1970–
    Desert / by Sheila Rivera.
       p.   cm. — (First step nonfiction)
    Includes index.
    ISBN: 0–8225–2597–6 (lib. bdg. : alk. paper)
    1. Deserts—Juvenile literature. 2. Desert animals—Juvenile literature. 3. Desert plants—Juvenile literature. I. Title. II. Series.
    QH88.R57 2005
    578.754—dc22   2004020790

Manufactured in the United States of America
1 2 3 4 5 6 – DP – 10 09 08 07 06 05